I0502434

TABLE OF CONTENTS:

ABOUT THE BOOK:

STEM is an acronym for Science, Technology, Engineering, and Mathematics. Recent shifts in education have favored these subjects, primarily because we have a shortage of workforce in these particular areas, which is really quite sad for a number of reasons. These can be some of the most interesting things to study in school, provided they are taught in a fun, interesting, and hands-on fashion. They also lead to some of the best-paying technical jobs, too!

All of the labs within promote these 4 fields, with an emphasis on holidays and seasons in the themes of the projects for this 4th collection of the series. It works as a stand-alone product or with the other volumes in the series.

Many of the details for each project have deliberately been left vague. It is important for students to design and work toward a goal, rather than have the process spelled out for them. This is a major element of the engineering-design process. However, since these are designed for an educational setting, optional grading suggestions have been added. For fun and homeschool use, discard these recommendations or make up your own grading system to suit your purpose.

This particular volume is an add-on and follow-up to 50 Holiday STEM Labs. It can be used with it or by itself as a stand-alone volume.

COPYRIGHT & USAGE:

CONTENTS: Mission Listing - Page 1

CONTENTS: Mission Listing - Page 2

CONTENTS: Mission Categories Listing

50 MORE HOLIDAY STEM LABS

Each of the new science labs in **50 MORE HOLIDAY STEM LABS** has the following:

- A snappy **Title**

- A **Brief Description** of the task to be completed

- General **Mission Rules**, suggestions, limitations, and requirements of the task

- Suggested **Materials Lists** for the project

- **Graded Assignment Suggestions** for Journaling and other associated projects

- **Category Tags** at the bottom to help you find projects with similar skills

CHRISTMAS

PROJECTS

CHRISTMAS: Christmas Lights

Christmas Lights are an instantly-recognizable symbol of winter and the holidays. Make your own fancy Christmas lights!

MISSION RULES AND RESTRICTIONS:

1. Create a Christmas light display without using string lights. This might include:

   A. a Kaleidoscope type design placed in front of light panels or flashlights

   B. a light box with holes in it that let out different colors in Christmas shapes

   C. an LED device

   D. a mixture of light sources arranged in a Christmas display

MATERIALS:

Materials may include:

- Clear & colored plastic sheets

- Markers

- boxes, colored paper, wrapping paper

- Stencils

- Scissors, glue, tape

- Glitter, Reflective decorations, rhinestones, and other shiny materials.

- Flashlights and/or panel lights, like a portable black light or fluorescent light

GRADED ASSIGNMENT SUGGESTIONS:

Grading suggestions include:

- a write-up of ideas, including blueprints

- grade results based upon if the device actually works, and, if so, how successfully?

- a reflection based on the experiences

- a project on lights, LEDs, light bulbs, the history of Christmas Lights, etc...

CATEGORIES: Lights, Reflection, Refraction

CHRISTMAS: Ice Boats

Ice boats are an awesome symbol of winter. Speeding across frozen lakes on metal runners, they can zip at very high speeds! Build your own sailboat on runners to catch the wind! Race your opponents.

MISSION RULES AND RESTRICTIONS:

1. Design an ice boat from scratch. Research might be required.

2. Test and retest your design on the flat surface under the power of wind.

3. Redesign as necessary to either:

   A. Have the boat that travels the farthest in the same conditions, or

   B. Have the boat that travels a specific distance faster than all other boats.

MATERIALS:

Materials may include:

- A flat table, tile floor, or smooth surface

- A floor fan/box fan

- Plastic straws

- Popsicle sticks

- Tape or glue

- Paper Clips

- Card stock or note cards

- paper, cloth, or coffee filters

- String

GRADED ASSIGNMENT SUGGESTIONS:

Grading suggestions include:

- a write-up of ideas, including blueprints

- grade results based upon if the device actually works, and, if so, how successfully?

- a reflection based on the experiences

- a project on ice boats

CATEGORIES: Boats, Distance, Friction, Speed, Wind

CHRISTMAS: Ice Fishing

Ice Fishing is another great Winter hobby! Make a

MISSION RULES AND RESTRICTIONS:

1. Make a fishing rod with a hook and/or magnet on the end.

2. Dip your hook in through the ice (sheet with 6" hole in it)

3. Try to snag or catch the fish hanging out beneath it.

4. Redesign to make your fishing pole as accurate and quick as it can be.

TEACHER'S NOTE: Beneath the sheet of 'ice' you should have half a dozen fish suspended at different depths or heights. This may work best if you place the board between two tables about 3 feet off the ground. Fish should be lightly attached with tape, velcro, or string and tacks near the edge of the fishing hole. By using different lengths of line, the fish can be suspended at different heights. Students should try to 'catch' as many magnetic fish or fish with loops near their mouths as they can in a set time.

MATERIALS:

Materials may include:

- A cardboard sheet, styrofoam board, or piece of plywood with a 6 inch hole in it.

- Magnets

- Paper clips

- Dowel rods

- thread spools

- string or fishing line

- Some sort of fish or target to hook/catch

GRADED ASSIGNMENT SUGGESTIONS:

Grading suggestions include:

- a write-up of ideas, including blueprints

- grade results based upon if the device actually works, and, if so, how successfully?

- a reflection based on the experiences

- a project on ice fishing

CATEGORIES: Devices, Engineering, Fishing, Ice

CHRISTMAS: Igloos

Can it get any colder than the Arctic? Build an igloo to survive frigid conditions!

MISSION RULES AND RESTRICTIONS:

1. Design a carton or igloo that you can place a small cup of water in.

2. Place it in the freezer.

3. Take it out and check periodically, perhaps every 15-30 minutes. The goal is to build an insulated box that will prevent freezing for as long as possible.

4. Return the igloo to the freezer and continue to check until it is frozen.

MATERIALS:

Materials may include:

• a freezer

• thermometers

• foam, styrofoam, packing peanuts, and other packing materials.

• a box or carton

• water

• small cup

GRADED ASSIGNMENT SUGGESTIONS:

Grading suggestions include:

• a write-up of ideas, including blueprints

• grade results based upon if the device actually works, and, if so, how successfully?

• a reflection based on the experiences

• a project on Inuit or Laplander culture or another native culture that live in cold regions.

CATEGORIES: Building, Engineering, Insulators, Temperature, Water

CHRISTMAS: Out of Control Snowball

Ever see those cartoons where a snowball rolls down a hill, growing larger and larger as it goes? That's your job! Build a snowball that will collect materials as it rolls down a slope.

MISSION RULES AND RESTRICTIONS:

1. Instructors (possibly with student help) will want to design a ramp for rolling projects down.

2. Design a snowball that will roll down a slope. As it rolls, it should pick up as many materials as possible.

3. The snowballs should be measured for weight/mass before and after the run to determine an increase in weight.

4. Multiple runs may be required. Between runs, resetting of adhesives or other materials may be required as well.

MATERIALS:

Materials may include:

- paper, puff balls, bubble wrap, aluminum foil, or other base materials for the snowballs

- glue, a variety of tapes, and other adhesives

- magnets

- scales to measure weight

- a ramp, possibly built from wood, construction paper, card stock, or cardboard.

- ramp litter materials like: cereal, sawdust, pretzels, paper clips, etc...

GRADED ASSIGNMENT SUGGESTIONS:

Grading suggestions include:

- a write-up of ideas, including blueprints

- grade results based upon if the device actually works, and, if so, how successfully?

- a reflection based on the experiences

- a project on adhesives.

- a project on sledding, tobogganing, or other downhill sports.

CATEGORIES: Collecting Materials, Gravity, Mass, Weight

CHRISTMAS: Ski Jumps

There are few sports as thrilling as a downhill ski jump. Design a ski jumper that can fly off of a custom ramp!

MISSION RULES AND RESTRICTIONS:

1. Instructors (possibly with student help) will want to design a ramp for jumping projects off of.

    A. The ramp will likely be around 3-6 foot long, usually curved upward at the end.
    B. It may have to be anchored onto a table, the wall, or other furniture.
    C. The ramp should end with a 2-3 foot drop to allow for longer jumps and more distance.

2. With the provided materials, students will design a project that can either fly as far as possible from the end of the jump or get the longest hang time.

3. Optionally, you may want to add a skier, like cotton balls and toothpicks. The skier should not fall off or land upside down.

MATERIALS:

Materials may include:

• note cards

• toothpicks

• popsicle sticks

• tape and glue

• tin foil

• cotton balls

• a ramp, possibly built from wood, construction paper, card stock, or cardboard.

• measuring tape

GRADED ASSIGNMENT SUGGESTIONS:

Grading suggestions include:

• a write-up of ideas, including blueprints

• grade results based upon if the device actually works, and, if so, how successfully?

• a reflection based on the experiences

• a project on the physics of ski jumping and other gliding or unpowered human flying activities.

CATEGORIES: Distance, Flight, Measurement, Winter Sports

CHRISTMAS: Sleigh Races

Create a sleigh that can hold a bag of toys and outrace its competition!

MISSION RULES AND RESTRICTIONS:

1. Create a small cart or sleigh no larger that 6 inches in any dimension.

2. The sleigh must have a spot to put a bag of weights, such as a bag of rice, coins, or sand.

3. The weight should not fall out of the project on the way to the end of the track.

4. Two different challenges can occur:

   A. How much weight can it carry down to the end of the track without collapsing?

   B. How fast can the sleigh carry a set amount of weight to the end of the track.

MATERIALS:

Materials may include:

- Small balloon or sock of rice or other weight.

- Plastic straws

- Popsicle sticks

- Tape or glue

- Paper Clips

- Card stock or note cards

- A track, ramp, or inclined table

GRADED ASSIGNMENT SUGGESTIONS:

Grading suggestions include:

- a write-up of ideas, including blueprints

- grade results based upon if the device actually works, and, if so, how successfully?

- a reflection based on the experiences

- a project on sleds and sleighs

CATEGORIES: Cars, Friction, Speed, Tracks, Time, Weight

CHRISTMAS: Snow Plows

During winter, you might just find yourself shoveling some snow, or plowing it if there is a lot of the white fluffy stuff. In this project, you need to build a snow shovel out of the provided materials!

MISSION RULES AND RESTRICTIONS:

1. Design a shovel that can scoop up snow faster than the other teams.

2. Teams will only get a set amount of note cards, tape, and plastic straws.

3. Within the time limits, build, test, and redesign your project.

4. Students will have to fill a cup with snow in the allotted time. Faster teams win.

MATERIALS:

Materials may include:

- powdered mash potatoes, shredded white paper, or other snow substitutes

- tape and glue

- plastic straws

- note cards

- cups or buckets to gather snow in

GRADED ASSIGNMENT SUGGESTIONS:

Grading suggestions include:

- a write-up of ideas, including blueprints

- grade results based upon if the device actually works, and, if so, how successfully?

- a reflection based on the experiences

- a project on levers and simple machines.

CATEGORIES: Machines, Task Completion, Time, Tools, Weather

EASTER

PROJECTS

EASTER: All Your Eggs in One Basket

Eggs are great, but broken ones in the carton are terrible! What a sad waste... Create a better carton that can protect eggs from breakage.

MISSION RULES AND RESTRICTIONS:

1. Create a carton that can hold a specified number of eggs, possibly 1, 2, 4, or 6.

2. The carton should protect all the eggs from breakage during a series of tests and real-life simulations, such as:

   A. short drop

   B. Getting mashed into a bag of groceries

   C. Regular shaking and jostling

   D. Whatever else your class determines

3. Test and redesign as necessary.

MATERIALS:

Materials may include:

- Eggs (real ones most likely)

- plastic drop cloths or easy-to-clean area

- cardboard, fabric, foam, styrofoam, cotton balls, etc...

- boxes, cups, and other containers

- tape, glue, string, rubber bands, etc...

GRADED ASSIGNMENT SUGGESTIONS:

Grading suggestions include:

- a write-up of ideas, including blueprints

- grade results based upon if the device actually works, and, if so, how successfully?

- a reflection based on the experiences

- a project about package design

CATEGORIES: Eggs, Engineering, Design, Packaging, Protection

EASTER: Bobbing for Eggs

It's not just for Halloween! Your job is to build the best tool to get Easter eggs out of a tub of water, leaving the water behind if possible. Speed is the key!

MISSION RULES AND RESTRICTIONS:

1. Build a device to pick a variety of sizes of eggs out of a tub of water.

2. Eggs might be submerged or floating.

3. Use only the provided materials.

4. Speed may be required. You might be timed!

5. You may not use your hands directly on the eggs. Only your tool may touch the egg, and your hands cannot get wet.

MATERIALS:

Materials may include:

- tub of water

- plastic eggs

- weights to submerge eggs (optional)

- plastic straws

- rubber bands

- toothpicks

- popsicle sticks

- plastic spoons

GRADED ASSIGNMENT SUGGESTIONS:

Grading suggestions include:

- a write-up of ideas, including blueprints

- grade results based upon if the device actually works, and, if so, how successfully?

- a reflection based on the experiences

- a project about buoyancy or eggs.

CATEGORIES: Accuracy, Buoyancy, Eggs, Grabbers, Time

EASTER: Bunny Hoppers

Bunnies are a pretty common sight around Easter. Create your own hopping rabbit device! How far or high can it jump?

MISSION RULES AND RESTRICTIONS:

1. Create a small model of a rabbit that can jump when activated.

2. The rabbit should not be more than 6 inches in any dimension.

3. The rabbit should activate or hop when pushed or triggered.

4. Measure the distance or height of the leap. One rabbit can be made for each purpose, if so desired.

5. Take an average measurement of the rabbit's jump by doing 3 or more tests.

6. Design and redesign as necessary.

MATERIALS:

Materials may include:

• rubber bands

• popsicle sticks

• plastic straws

• string

• tape

• glue

• paper

GRADED ASSIGNMENT SUGGESTIONS:

Grading suggestions include:

• a write-up of ideas, including blueprints

• grade results based upon if the device actually works, and, if so, how successfully?

• a reflection based on the experiences

• a project about rabbits

CATEGORIES: Animals, Distance, Height, Measurement, Models, Movement

EASTER: Egg Chute

Create an egg chute to collect eggs from as far across a room as possible.

MISSION RULES AND RESTRICTIONS:

1. Make as long of a chute as possible to transfer eggs across a room into a box or cup.

2. The chute should be free-standing or attached to furniture/walls as approved.

3. Distance will be measured in a straight line from the entry point to the collection container.

4. Restarts or pushes to help the eggs along will be measured from that point, not the start point.

5. Multiple eggs will be tested. An average distance traveled without assistance will be used for your final score.

MATERIALS:

Materials may include:

• Tape, String, Glue,

• Cardboard tubes

• Foam pipe insulation

• Card Stock

• Plastic Easter Eggs, or chocolate eggs

GRADED ASSIGNMENT SUGGESTIONS:

Grading suggestions include:

• a write-up of ideas, including blueprints

• grade results based upon if the device actually works, and, if so, how successfully?

• a reflection based on the experiences

• a project about rabbits

CATEGORIES: Accuracy, Chutes, Distance, Eggs, Measurement, Tracks

EASTER: Egg-a-Pults

Easter Eggs are everywhere, even flying! Make a device that can fling a plastic Easter egg as far as possible.

MISSION RULES AND RESTRICTIONS:

1. Create a device that can throw a plastic egg as far as possible.

2. The device should have a triggering mechanism or some way to toss the egg on command and be reloaded.

3. The device should not be more than 1 foot in any dimension when at rest.

4. Measure the distance or height of the egg toss.

5. Take an average measurement of the egg's flight by doing 3 or more tests.

6. Design and redesign as necessary.

MATERIALS:

Materials may include:

• rubber bands

• popsicle sticks, dowel rods, or sticks

• plastic straws

• string or elastic cord

• tape

• glue

• paper

• plastic cups

• plastic easter eggs

GRADED ASSIGNMENT SUGGESTIONS:

Grading suggestions include:

• a write-up of ideas, including blueprints

• grade results based upon if the device actually works, and, if so, how successfully?

• a reflection based on the experiences

• a project about siege weapons or eggs

CATEGORIES: Distance, Measurement, Siege, Throwers

EASTER: Egg-athlon

Egg hunts don't have to be easy. Go through or help create a STEM challenge course where you must complete challenges to collect eggs.

MISSION RULES AND RESTRICTIONS:

1. Create a series of tasks, or help create a single STEM task to be put with other tasks created by peers, as part of an obstacle course Easter Egg Hunt.

2. Your task should be something that can be completed in just a few minutes. Clear rules and expectations should be explained.

3. Upon completion of a task, eggs/rewards should be offered, and students will move to the next station.

4. All stations should be completed or attempted within the time limits, when possible.

5. Voting may occur for best/most fun challenge.

MATERIALS:

Materials may be pretty intensive for this project. They will vary greatly, but will include:

- Supplies to create a map or route for challengers to follow

- Supplies for students to create a variety of STEM challenges

GRADED ASSIGNMENT SUGGESTIONS:

Grading suggestions include:

- a write-up of ideas, including blueprints

- grade results based upon if the challenge is actually designed properly according to the requirements.

- a reflection based on the experiences

- a project about multiple part sporting events or challenges

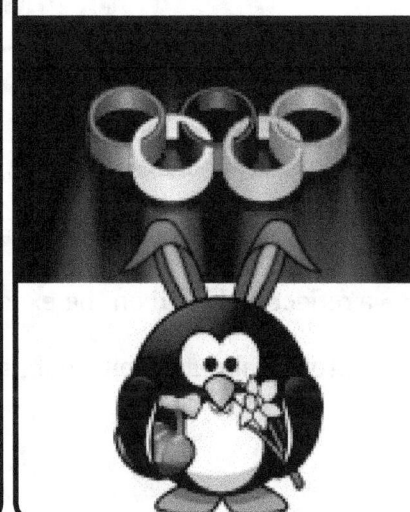

CATEGORIES: Multiple Events, Student-Created Challenges

EASTER: Rabbitats

You want the Easter Bunny to stop by, right? Well, you need to make a welcoming habitat for the rabbit, a Rabbitat.

MISSION RULES AND RESTRICTIONS:

1. Design a miniature habitat for an Easter Bunny.

2. Make sure to include all the things they need or like, such as:

   A. Eggs

   B. Water

   C. Grass

   D. Flowers

MATERIALS:

Materials may include:

- Easter Grass

- Packing Materials

- Boxes

- Colorful decorations

- Eggs

- Flowers

- Other Easter or rabbit-themed decorations

GRADED ASSIGNMENT SUGGESTIONS:

Grading suggestions include:

- a write-up of ideas, including blueprints

- have students vote for their favorite and/or most spirited

- a reflection based on the experiences

- a project about rabbits, habitats, or the needs of small rodents

CATEGORIES: Animals, Environments, Habitats, Life Science

EASTER: Stylin' Eggs

Dipping eggs to color them is so last season. It's time to decorate your eggs in a better way. Presentation is <u>everything</u>.

MISSION RULES AND RESTRICTIONS:

1. Create a device or tool to help you color your eggs in a different way. You might wish to consider:

   A. Blown Paint or Air Brushing

   B. Decorations added to eggs

   C. Paint, Markers, or other dye sources

   D. Protectants to prevent color on some parts of the eggs

2. Test your device on multiple eggs.

3. Compare to other techniques to determine which makes the most attractive/stylish eggs.

MATERIALS:

Materials may include:

- Plastic straws

- food coloring, washable markers, or other safe dyes (including foods)

- plastic straws

- paint brushes

- drop clothes, newspapers, painting aprons/shirts

- containers & water

GRADED ASSIGNMENT SUGGESTIONS:

Grading suggestions include:

- a write-up of ideas, including blueprints

- grade results based upon if the device actually works, and, if so, how successfully?

- a reflection based on the experiences

- a project about dyes and coloring fabrics and other materials

CATEGORIES: Art, Eggs, Paint, Task Completion

HALLOWEEN

PROJECTS

HALLOWEEN: Bobbing for Apples

A favorite Halloween game of many people is Bobbing for Apples. In this mission, you will devise a tool to scoop up or pick up an apple or apple replica from a pot or bucket of water.

MISSION RULES AND RESTRICTIONS:

1. Design a tool from available or scavenged materials. Suggestions are:

   A. a scooper

   B. a pick

   C. tongs

   D. Arm with moving joints

2. Test to see how many apples your tool can pull from the water in a set amount of time. Your teacher will determine these conditions, but may be up to 10 apples within 10 seconds, or something like that.

GRADED ASSIGNMENT SUGGESTIONS:

Grading suggestions include:

• a write-up of ideas, including blueprints

• grade results based upon if the device actually works, and, if so, how successfully?

• a reflection based on the experiences

• a project about levers and simple machines

MATERIALS:

Materials may include:

• real apples, ping pong balls, red rubber balls, or something to approximate an apple

• rubber bands

• plastic straws

• toothpicks

• tape

• bamboo skewers

• fishing line

• other scavenged office and household supplies

• stopwatch

CATEGORIES: Accuracy, Simple Machines, Task Completion, Tools

HALLOWEEN: Candy Sweepers

Witches and brooms are great Halloween symbols. Why not make a broom to sweep up all of your candy plunder?

MISSION RULES AND RESTRICTIONS:

1. Design a broom that can sweep candy into a bucket or bag as fast as possible without missing any pieces.

2. Test and redesign the broom to get the best time possible.

3. The project may NOT just be a simple shovel or scooper, but must have individual bristles and resemble an actual broom.

4. Measure the time taken and the accuracy collecting different shapes, weights, and sizes of candy.

MATERIALS:

Materials may include:

• plastic straws

• pipe cleaners

• hay, grasses, or twigs

• dowel rods

• tape and glue

• string or wire

• bamboo skewers

• a pile of candy of assorted shapes, sizes, and weights

• buckets or bags to sweep into

• stopwatch

GRADED ASSIGNMENT SUGGESTIONS:

Grading suggestions include:

• a write-up of ideas, including blueprints

• grade results based upon if the device actually works, and, if so, how successfully?

• a reflection based on the experiences

• a research project on the origins of a household tool or the international/cultural varieties of that tool

CATEGORIES: Task Completion, Time, Tools, Work

HALLOWEEN: Egging Houses

The trick side of 'trick or treat' is often forgotten for sweet candy and fun costumes. Why not try causing a little mischief? Design a project that throws eggs (real or fake) or toilet paper?

MISSION RULES AND RESTRICTIONS:

1. Design a device that can throw either eggs (real or fake substitutes) or toilet paper as far or accurately as possible.

2. Test and redesign the device to get the best distance and/or accuracy possible.

3. The project should be made from available materials, scavenged from home and school or provided by the instructor.

4. Measure the distances and or accuracies of the device.

OPTION: Set up cardboard painted houses as targets.

MATERIALS:

Materials may include:

• rubber bands

• glue, duct tape

• wire, pipe cleaners

• wood, popsicle sticks, dowel rods, bamboo skewers, etc...

• eggs or ping pong balls, or the optional toilet paper

• targets, like cardboard boxes or hula hoops

• measuring tape

GRADED ASSIGNMENT SUGGESTIONS:

Grading suggestions include:

• a write-up of ideas, including blueprints

• grade results based upon if the device actually works, and, if so, how successfully?

• a reflection based on the experiences

• a research project on early siege weapons, like catapults and trebuchets

CATEGORIES: Accuracy, Distance, Machines, Simple Machines

HALLOWEEN: Floating Ghosts

What is Halloween without a spooky ghost or seven? Make a ghost that floats to the ground as slowly as possible.

MISSION RULES AND RESTRICTIONS:

1. Design a floating ghost from the provided materials.

2. Test and redesign the project to get the best hang time (the slowest fall).

3. Measure the time it takes for your project to fall from a specified height, such as:

   A. a person standing on top of a chair or table

   B. a person on the top of a ladder

MATERIALS:

Materials may include:

• ladder or chair

• coffee filters, muffin tray liners, tissue paper, newspaper, plastic wrap, and other pieces of light material cut into about the same sizes.

• streamers or ribbons

• string

• glue or tape

• scissors

• markers to draw ghost faces

• stopwatch

GRADED ASSIGNMENT SUGGESTIONS:

Grading suggestions include:

• a write-up of ideas, including blueprints

• grade results based upon if the device actually works, and, if so, how successfully?

• a reflection based on the experiences

• a project on bats

CATEGORIES: Flight, Hang Time, Time

HALLOWEEN: Monster Mashers

Monsters are about, and they like to smash stuff. Get them busy smashing marshmallows. Build a project that can flatten a marshmallow as flat as possible without you physically touching it.

MISSION RULES AND RESTRICTIONS:

1. Design and build a project that can crush a marshmallow without you having to touch it.

2. You may not simply drop something on the marshmallow. Sorry!

3. You will get up to 3 trials to determine how flat you can make a marshmallow.

OPTION: Check the original size, the crushed size, and the re-inflated size 10 seconds after the marshmallow has been released from the project.

OPTION: You could also measure marshmallow volume before and after.

MATERIALS:

Materials may include:

• marshmallows (mini, regular, or jumbo)

• ruler

• plastic straws and spoons

• bamboo skewers, plastic straws, and pipe cleaners

• tape, glue

• string, wire

• other scavenged materials

GRADED ASSIGNMENT SUGGESTIONS:

Grading suggestions include:

• a write-up of ideas, including blueprints

• grade results based upon if the device actually works, and, if so, how successfully?

• a reflection based on Halloween experiences or on the project

• a project on marshmallows

CATEGORIES: Crushing, Device, Size, Smashing

HALLOWEEN: Straw Mazes

Straw Mazes can be a lot of fun. Run into them with friends and find your way out! Your mission is to develop a straw maze, but it won't be in a hayfield - it will be made from plastic straws!

MISSION RULES AND RESTRICTIONS:

1. Design and build a maze for a marble or small ball using the provided materials.

2. You should be able to pick up the maze and tilt it back and forth to maneuver the ball from a designated start to a finish point.

3. The project can't be a simple straight line or less than three turns.

4. Bonus points for more complicated mazes!

MATERIALS:

Materials may include:

- cardboard and/or sheets of styrofoam

- tape, glue, scissors

- plastic straws (LOTS)

- toothpicks (optional substitute for straws)

- marbles, bouncy balls, or ping pong balls

GRADED ASSIGNMENT SUGGESTIONS:

Grading suggestions include:

- a write-up of ideas, including blueprints

- grade results based upon if the device actually works, and, if so, how successfully?

- a reflection based on Halloween experiences or on the project

- a project on mazes, the history of mazes, or straw mazes

- a photo collection/collage of straw mazes.

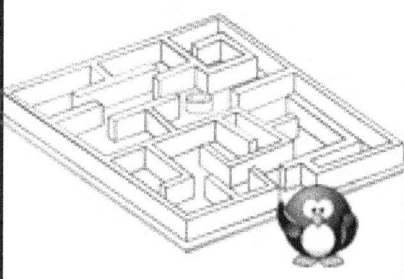

CATEGORIES: Light, Models, Plants

HALLOWEEN: Witches' Brew

A witch (or warlock) needs a great big black cauldron to brew up some potions in. Practice making a cauldron bubble over with the wrong mix of ingredients!

MISSION RULES AND RESTRICTIONS:

1. Design and build a device that can deliver pellets of baking soda (Alka Seltzer pieces) or powdered baking soda into a cauldron from as far back as possible.

2. Your device should be loaded from a specific spot, and it should deliver the baking soda or Alka Seltzer into the pot as efficiently as possible from as far away as possible.

3. The cauldron will be a stationary pot or bucket with vinegar , food coloring, and a little bit of dish soap for effect (just like the volcano project, but maybe green)

4. The most successful project from the farthest away wins!

GRADED ASSIGNMENT SUGGESTIONS:

Grading suggestions include:

• a write-up of ideas, including blueprints

• grade results based upon if the device actually works, and, if so, how successfully?

• a reflection based on Halloween experiences or on the project

• models of compounds, especially household chemicals and things like water

MATERIALS:

Materials may include:

• at least one bucket or plastic decorative cauldron

• vinegar, food coloring, and dish soap

• alka-seltzer or baking soda

• plastic straws and spoons

• bamboo skewers, plastic straws, and pipe cleaners

• tape, glue

• string, wire

• other scavenged materials

• measuring tape

CATEGORIES: Accuracy, Chemical Reactions, Chemistry, Distance

PATRIOTIC

PROJECTS

PATRIOTIC THEMES: 50 States

Create a model topographical map of the USA.

MISSION RULES AND RESTRICTIONS:

1. Build a scale model map of the USA or of a particular state.

2. Paint and color it according to topography and landforms.

3. Use little flags or markers to denote important locations, which might require a legend or site listing.

4. Add in bodies of water and neighboring countries, if applicable.

5. Decorate the map.

MATERIALS:

Materials may include:

- play dough, clay, salt dough, or any other workable medium

- cardboard or plastic trays to build the map on.

- paint

- gravel or sand

- glue

- other decorations

GRADED ASSIGNMENT SUGGESTIONS:

Grading suggestions include:

- a write-up of ideas, including blueprints

- grade results based upon the success of the project

- a reflection based on the experiences

- a project on American expansion, a particular state, or American territory

CATEGORIES: Clay, Maps, Models, Patriotic, USA

PATRIOTIC THEMES: Crossing the Delaware

Design a boat that can cross a rough, icy river, carrying its crew to the other side.

MISSION RULES AND RESTRICTIONS:

1. Design a boat that can carry 3 soldiers across the windy, icy river (ice cubes floating in a tray or sink with a fan blowing across it). Build it with the provided materials, which may be limited.

2. The boat must make it across the river without sinking.

3. The boat should get all 3 passengers across the river without having them fall out. Try to keep them as dry as possible, too!

4. Test and redesign for success. Testing opportunities might be limited.

MATERIALS:

Materials may include:

• water trough or sink

• ice cubes

• a box fan or other small fan

• army men or small plastic characters

• tin foil or wax paper

• popsicle sticks

• glue and tape

• index cards

GRADED ASSIGNMENT SUGGESTIONS:

Grading suggestions include:

• a write-up of ideas, including blueprints

• grade results based upon if the device actually works, and, if so, how successfully?

• a reflection based on the experiences

• a project on an American military operation, particularly in the Revolutionary War

CATEGORIES: Boats, Buoyancy, Task Completion, Vehicles, Water

PATRIOTIC THEMES: Fireworks in a Jar

Create a fireworks display in a jar with some creative food coloring and other supplies!

MISSION RULES AND RESTRICTIONS:

1.   Create a bright jar full of decorations in either:

   A. American colors of red, white, and blue theme

   B. Colorful fireworks theme

MATERIALS:

Materials may include:

- Vegetable oil
- Water
- Food Coloring
- Glitter/sparkly stuff
- Jars and containers
- Aqua beads

GRADED ASSIGNMENT SUGGESTIONS:

Grading suggestions include:

- a write-up of ideas, including blueprints
- grade results based upon if the device actually works, and, if so, how successfully?
- a reflection based on the experiences
- a project on fireworks
- a project on mixtures, suspensions, solutions, and solubility

CATEGORIES: Mixtures, Solutions, Suspensions

PATRIOTIC THEMES: Presidential Memorial

Create a monument or memorial to a past or present president of your choice.

MISSION RULES AND RESTRICTIONS:

1. Research a president as well as any existing presidential monuments to get ideas.

2. Scavenge for materials to make a memorial.

3. Build the project. Decorate it.

4. Present the project alongside a short report on a president.

MATERIALS:

Materials may include:

- Boxes, Cups, Containers, and other packaging

- Glue, Tape, String, and Rubber Bands

- Clay, Rocks, Play-Dough, and other modeling mediums

- Paint, Markers

- Colored Paper, Cardboard

GRADED ASSIGNMENT SUGGESTIONS:

Grading suggestions include:

- a write-up of ideas, including blueprints

- grade results based upon the success of the project, especially how recognizable it is.

- a reflection based on the experiences

- a project on a president.

CATEGORIES: Culture, Models, Presidents, Social Studies, USA

PATRIOTIC THEMES: Recycled Monuments

America is filled with amazing examples of architecture and natural beauty. Make a recreation of one such example with recycled and scavenged materials.

MISSION RULES AND RESTRICTIONS:

1. Research natural formations and/or examples of famous architecture in America.

2. Scavenge for materials to make a model recreation.

3. Build the project. Decorate it for accuracy.

4. Present the project alongside a picture of the actual thing.

MATERIALS:

Materials may include:

- Boxes, Cups, Containers, and other packaging

- Glue, Tape, String, and Rubber Bands

- Clay, Rocks, Play-Dough, and other modeling mediums

- Paint, Markers

- Colored Paper, Cardboard

GRADED ASSIGNMENT SUGGESTIONS:

Grading suggestions include:

- a write-up of ideas, including blueprints

- grade results based upon the success of the project, especially how recognizable it is.

- a reflection based on the experiences

- a project on national monuments or natural formations.

CATEGORIES: Culture, Geology, Models, Social Studies

PATRIOTIC THEMES: Red, White, and Blue

Make a patriotic craft or design a patriotic activity focusing on our country's 3 colors: red, white, and blue.

MISSION RULES AND RESTRICTIONS:

1. Create a project based on the 3 colors of the American flag: red, white, and blue.

2. Your project may be just about anything, created from any available and scavenged materials at home and school, provided that it is primarily red, white, and blue, and that it has some relationship to our country.

MATERIALS:

Materials will greatly vary for this project, depending on the type of project chosen. Generally, offer a lot of paint, markers, and colored items, as well as lots of arts and craft materials.

GRADED ASSIGNMENT SUGGESTIONS:

Grading suggestions include:

• a write-up of ideas, including blueprints

• grade results based upon the success of the project

• a reflection based on the experiences

• a project on the American Flag or other American symbols.

CATEGORIES: Art, Design, Patriotic

ST. PATRICK'S DAY

PROJECTS

ST. PATRICK'S DAY: Green Houses

Make and decorate an Irish-themed mini greenhouse.

MISSION RULES AND RESTRICTIONS:

1. Scavenge for recycled or reusable materials that can be made into a greenhouse.

2. The greenhouse should be enclosed, likely with a transparent or translucent film to trap heat and moisture.

3. Make sure there is some sort of drainage and/or irrigation.

4. Also find a growing medium, which might include gravel or perlite at the bottom.

5. If possible, include clovers from your yard or clover seeds. Small cactuses and other decorative and miniature plants are great, too!

MATERIALS:

Materials may include:

• plastic jugs

• pipes, straws, or tubes

• PVC pipes

• wood

• plastic sheeting, plastic wrap, etc...

• gravel, soil, peat, coconut fiber, perlite, potting soil, etc...

• water

• seeds

• miniature plants or plants from outside

GRADED ASSIGNMENT SUGGESTIONS:

Grading suggestions include:

• a write-up of ideas, including blueprints

• grade results based upon if the device actually works, and, if so, how successfully?

• a reflection based on the experiences

• a project about raised bed gardens, micro gardens, vertical gardening, and other gardening techniques.

CATEGORIES: Design, Engineering, Gardening, Life Cycles, Plants

ST. PATRICK'S DAY: Green Machine

Make a device that can turn a cup of water or beverage green.

MISSION RULES AND RESTRICTIONS:

1. Design a project that can add a drop or drops of food coloring to a beverage or cup of water.

2. The project should have a base or platform where a cup or beverage can be set. Activating your project should dispense a drop or several drops of food coloring into the water or beverage.

3. The device should be able to accurately dispense the same amount of food coloring with each activation.

4. Design, test, and redesign as necessary.

MATERIALS:

Materials may include:

• Water or beverages

• Cups

• Food Coloring

• Tubing

• Eyedroppers

• Popsicle sticks

• Tape, Glue, Rubber Bands

GRADED ASSIGNMENT SUGGESTIONS:

Grading suggestions include:

• a write-up of ideas, including blueprints

• grade results based upon if the device actually works, and, if so, how successfully?

• a reflection based on the experiences

• a project on drink machines, dispensers, and vending machines

CATEGORIES: Devices, Machines, Measurement, Water

ST. PATRICK'S DAY: Irish Jig

Make a Leprechaun dance!

MISSION RULES AND RESTRICTIONS:

1. Watch at least one clip or video of the Irish jig being performed.

2. Create a small character with moveable arms and legs.

3. Build the character in such a way that it can be manipulated and moved, so that it looks as if it is dancing.

4. Practice and work out a dance routine, especially if it can be put to an Irish or Celtic tune.

MATERIALS:

Materials may include:

• cardboard

• strings

• rubber bands

• glue and tape

• paint, markers, etc...

• popsicle sticks, skewers

• decorations

GRADED ASSIGNMENT SUGGESTIONS:

Grading suggestions include:

• a write-up of ideas, including blueprints

• grade results based upon if the device actually works, and, if so, how successfully?

• a reflection based on the experiences

• a project on folk dances, particularly the Irish Jig.

CATEGORIES: Culture, Dancing, Devices, Movement

ST. PATRICK'S DAY: Magic Jars

Make a green, Irish-themed magic relaxing jar!

MISSION RULES AND RESTRICTIONS:

1. Add clear school glue (lots) and glycerin (a little) to hot water. Stir until well-mixed.

2. Add decorative materials.

3. Top off the container with more water until full.

4. Seal tightly.

5. Shake and enjoy!

6. Make sure to collect materials that fit the St. Patrick's day theme.

MATERIALS:

Materials may include:

- Food coloring

- Glitter

- Sparkly stuff, especially gold and copper colored metal bits

- Shamrock-shaped beads or metallic bits

- Jars or small clear pill containers

- clear school glue

- glycerin

- hot water

GRADED ASSIGNMENT SUGGESTIONS:

Grading suggestions include:

- a write-up of ideas, including blueprints

- grade results based upon if the device actually works, and, if so, how successfully?

- a reflection based on the experiences

- a project on mixtures, suspensions, and solutions, as well as solubility

CATEGORIES: Mixtures, Solubility, Suspensions, Water

ST. PATRICK'S DAY: Money Counters

Create a device that can count how many coins you collect.

MISSION RULES AND RESTRICTIONS:

1. Design a project that can count coins or small objects.

2. The project should have a hopper or cup where objects are poured in. After, the objects should drain into a collection cup or reservoir. As the objects move from start to finish, they should be counted by mechanisms in the project.

3. The device should be able to accurately count how many objects are inside it. The number of objects should be displayed and easily readable.

4. Design, test, and redesign as necessary.

MATERIALS:

Materials may include:

- Coins or something to count, such as jellybeans, chocolate coins, marbles, etc...

- cardboard

- string, rubber bands

- glue, tape

- scissors

- spools, cups

- markers

- building toys, such as gears and other machine type parts

GRADED ASSIGNMENT SUGGESTIONS:

Grading suggestions include:

- a write-up of ideas, including blueprints

- grade results based upon if the device actually works, and, if so, how successfully?

- a reflection based on the experiences

- a project on different coins of the world, especially on the sizes, shapes, and other decorations, such as holes or layers.

CATEGORIES: Coins, Counting, Devices, Machines

ST. PATRICK'S DAY: Rainbow Makers

Rainbows lead to the pots of gold! Make a device that can create a spray of water that creates a rainbow!

MISSION RULES AND RESTRICTIONS:

1. Design a project that can create a rainbow.

2. The project should have a trigger, button, or pedal that allows you to turn it on or start it.

3. The device should be able to create a rainbow in sunlight or other light that can be photographed.

4. Design, test, and redesign as necessary.

MATERIALS:

Materials may include:

• Water sprayers

• Containers

• Water

• Tubing

• Sunlight or light source

• Crystals or other refractive materials

GRADED ASSIGNMENT SUGGESTIONS:

Grading suggestions include:

• a write-up of ideas, including blueprints

• grade results based upon if the device actually works, and, if so, how successfully?

• a reflection based on the experiences

• a project on rainbows, light, refraction, or prisms

CATEGORIES: Light, Rainbows, Refraction, Water, Weather

ST. PATRICK'S DAY: Shamrock Stamper

Create a machine that can create a shamrock at the touch of a button or lever!

MISSION RULES AND RESTRICTIONS:

1. Design a project that can create a shamrock shape or print.

2. The project should have a trigger, button, or pedal that allows you to turn it on or start it.

3. The device should be able to create a shamrock-shaped cutout, stamp, or other mark on a piece of paper.

4. Design, test, and redesign as necessary.

MATERIALS:

Materials may include:

- Paint

- Markers

- Foam sheets

- Popsicle sticks, bamboo skewers

- plastic straws

- rubber bands, tape, glue

- scissors

GRADED ASSIGNMENT SUGGESTIONS:

Grading suggestions include:

- a write-up of ideas, including blueprints

- grade results based upon if the device actually works, and, if so, how successfully?

- a reflection based on the experiences

- a project on stamps, print-making, and other artistic media.

CATEGORIES: Devices, Engineering, Machines, Simple Machines

THANKSGIVING

PROJECTS

THANKSGIVING: Acorn Collectors

Acorns are a great fall theme. Squirrels are scurrying around and gathering them up to get ready for the long winter. Practice being a squirrel on your own by making an acorn collector device.

MISSION RULES AND RESTRICTIONS:

1. Design an acorn collector device from the provided materials.

2. You must be able to pick up as many acorns (or all of a set amount placed out for you) as quickly as possible.

3. You may not touch the acorns with your hands or any part of your body. Do not nudge them into or onto your device with your shoe or anything like that.

4. Collect the acorns as fast as possible into a cup, your partner's hands, or onto a table/designated location.

5. Compare times with other teams. Faster is better!

MATERIALS:

Materials may include:

• popsicle sticks

• rubber bands

• toothpicks

• tape

• string

• plastic straws

• something to use as acorns, such as: coin operated toy dispenser capsules, styrofoam balls, actual acorns, any other tree nut still in shells, ping pong balls...

GRADED ASSIGNMENT SUGGESTIONS:

Grading suggestions include:

• a write-up of ideas, including blueprints

• grade results based upon if the device actually works, and, if so, how successfully?

• a reflection based on the experiences

• a project on tree nuts and/or seed dispersal

• a project on squirrels

CATEGORIES: Accuracy, Design, Time, Tools

THANKSGIVING: Cornucopias

Cornucopias are a symbol of harvest and plenty. Fill up your own cornucopia by creating a launching device that can fling foods into your own cornucopia.

MISSION RULES AND RESTRICTIONS:

1. Design a device that can toss, fling, or throw 'food' into your basket.

2. Your device may not simply pinch or pick up materials.

3. You instructor will designate a minimum distance you must shoot from beyond.

4. Build and test the device. Redesign as needed for accuracy.

5. Given a set amount of time, see which team(s) can fill their cornucopias up the most with food.

MATERIALS:

Materials may include:

- woven baskets, mini cornucopias, or any other cornucopia substitute

- popsicle sticks, toothpicks, and/or short dowel rods

- rubber bands, string, or pipe cleaners

- tape, glue

- plastic straws

- string or fishing line

- food for the cornucopia, which might be toy plastic food, cheese puff balls, dry cereal, or even cotton puff balls in a variety of colors

GRADED ASSIGNMENT SUGGESTIONS:

Grading suggestions include:

- a write-up of ideas, including blueprints

- grade results based upon if the device actually works, and, if so, how successfully?

- a reflection based on the experiences

- a project about crops being harvested in fall.

- a project about harvest festivals around the world

CATEGORIES: Accuracy, Distance, Engineering, Physics, Throwers, Time

THANKSGIVING: Gravy Boats

Gravy is the universal sauce for Thanksgiving dinner. It goes on everything. In this project, you are to build a gravy boat, and that's not the special dish for pouring gravy. You will build an actual boat that holds as much gravy as possible without leaking or sinking.

MISSION RULES AND RESTRICTIONS:

1. Build a boat from a sheet of tin foil.

2. Create it with the largest volume possible, but also leak-free.

3. Test it to see how much of a volume of gravy it can hold.

4. The more it holds, the better.

MATERIALS:

Materials may include:

- Gravy, preferably the powdered kind you can make lots of cheaply and quickly

- tin foil

- graduated measuring cup to pour from

- tub or sink full of water

GRADED ASSIGNMENT SUGGESTIONS:

Grading suggestions include:

- a write-up of ideas, including blueprints

- grade results based upon if the device actually works, and, if so, how successfully?

- a reflection based on the experiences

- a project about calculating volume

CATEGORIES: Boats, Buoyancy, Volume

THANKSGIVING: Pilgrim's Progress

We wouldn't have had Thanksgiving without the Pilgrims. They needed to set up settlements in the New World. Can you build a model settlement?

MISSION RULES AND RESTRICTIONS:

1. Design an early settlement out of scavenged materials.

2. Try to include major buildings and important things like:

   A. Palisade Walls

   B. Town Hall, Church, Homes, Places of Business

   C. A port or docks

   D. Farms

   E. Trees and forests

MATERIALS:

Materials may include:

- popsicle sticks

- cardboard

- toothpicks

- clay

- paint, glue, tape

- twigs and sticks

- rocks or fish tank gravel

- other scavenged materials from home and school

GRADED ASSIGNMENT SUGGESTIONS:

Grading suggestions include:

- a write-up of ideas, including blueprints and plans

- grade results based upon if the project is completed.

- a reflection based on the experiences

- a project on historical early American settlements

- maps collection of early settlements

CATEGORIES: Architecture, Civil Engineering, Maps, Social Studies

THANKSGIVING: Pumpkin Pie Eaters

Everyone loves pumpkin pie (or at least I do). Sometimes, you just want to eat it from a long distance away. Partner up with someone and make tools to feed your partner some pumpkin pie!

MISSION RULES AND RESTRICTIONS:

1. Design a super long feeding device.

2. Your device should be able to scoop up pumpkin pie and feed it to a partner across the table.

3. You may not get up out of your seat. You should be able to reach the pie in the middle of the table AND reach across the table to feed your partner.

4. Take turns feeding each other. Try to to get pie everywhere.

MATERIALS:

Materials may include:

• pumpkin pie

• spoons

• popsicle sticks

• plastic straws

• string, tape, pipe cleaners, or rubber bands

• bamboo skewers

GRADED ASSIGNMENT SUGGESTIONS:

Grading suggestions include:

• a write-up of ideas, including blueprints

• grade results based upon if the device actually works, and, if so, how successfully?

• a reflection based on the experiences

• a project on pumpkins

• a project about pies

CATEGORIES: Accuracy, Food, Teamwork, Tools

THANKSGIVING: Scarecrows

Fall is a time of harvest. It's really hard to harvest if birds and other critters have stolen all of your crops! Design a scarecrow that makes noise or otherwise scares away birds.

MISSION RULES AND RESTRICTIONS:

1. Design a scarecrow.

2. Your scarecrow must somewhat resemble a person or traditional scarecrow, but it must also have at least 1 more way to scare away birds, other than just its humanoid appearance. Ideas are:

   A. Makes noise in the wind to scare away birds

   B. Flashes or shines to distract/scare away birds

   C. Moves to scare away birds

MATERIALS:

Materials may include:

- popsicle sticks, toothpicks, bamboo skewers, and/or short dowel rods

- rubber bands, string, or pipe cleaners

- tape, glue

- felt or fabrics

- straw

- string or fishing line

- tin cans, bottle caps, old CD's

GRADED ASSIGNMENT SUGGESTIONS:

Grading suggestions include:

- a write-up of ideas, including blueprints

- grade results based upon if the device actually works, and, if so, how successfully?

- a reflection based on the experiences

- a project about pest/varmint control in farms by scarecrows and other means.

CATEGORIES: Agriculture, Animal Behavior, Models, Movement, Sound

THANKSGIVING: Teepee Time

Thanksgiving would not have been what it was without the help of the Native Americans. In honor of that cooperation, design a teepee, wigwam, roundhouse, pueblo, sod home, or other shelter from the given materials.

MISSION RULES AND RESTRICTIONS:

1. Design a model of a Native American shelter from the given materials.

2. The shelter should have a designated entry point.

3. The shelter can be decorated, but the design is the important part.

MATERIALS:

Materials may include:

- popsicle sticks, toothpicks, bamboo skewers, and/or short dowel rods

- rubber bands, string, or pipe cleaners

- plastic straws

- tape, glue

- construction paper, coffee filters, or felt for coverings

- clay and straw

GRADED ASSIGNMENT SUGGESTIONS:

Grading suggestions include:

- a write-up of ideas, including blueprints

- grade results based upon if the device actually works, and, if so, how successfully?

- a reflection based on the experiences

- a project on different shelter types by region or tribe of Native Americans

CATEGORIES: Architecture, Cultural Studies, Engineering, Social Studies

VALENTINE'S DAY

PROJECTS

VALENTINE'S DAY: A Rose By Any Other Name

Create a replica of a rose.

MISSION RULES AND RESTRICTIONS:

1. Create the most realistic model of a rose possible.

2. Study actual roses or detailed pictures of them to get ideas.

3. Assemble and build the rose.

MATERIALS:

Materials may include:

- pipe cleaners
- tissue paper
- construction paper
- scissors
- markers
- paint
- glue
- tape
- origami paper
- modeling clay
- felt or thin cloth

GRADED ASSIGNMENT SUGGESTIONS:

Grading suggestions include:

- a write-up of ideas, including blueprints
- a project classifying flowers
- a project about the meanings of flowers

CATEGORIES: Design, Models, Plants

VALENTINE'S DAY: All Boxed Up

Create a heart-shaped box from scratch.

MISSION RULES AND RESTRICTIONS:

1. Design and create a heart-shaped box without using any template.

2. Your box should be 3D and should have a lid that fits on it.

3. Decorate it. Style counts!

4. NOTE: Ideally, the box will not just be a square box with a heart on top, but that might be suitable for younger students.

MATERIALS:

Materials may include:

• construction paper

• scissors

• tape and glue

• card stock

• cardboard

• ribbon

• wrapping paper

GRADED ASSIGNMENT SUGGESTIONS:

Grading suggestions include:

• a write-up of ideas, including blueprints

• a reflection based on the experiences

• create other 3D boxes in different shapes and styles, as well as nets to create other 3D shapes

CATEGORIES: Design, Geometry, Nets, Shapes

VALENTINE'S DAY: Box of Chocolates

Decorate your own box of chocolates.

MISSION RULES AND RESTRICTIONS:

1. Decorate and design your own box of chocolates.

2. Design and decorate the box first.

3. Cut and fit a candy tray to fit, or create your own tray with sections to keep the candies separate. You can also use mini muffin tray liners.

4. Decorate candies and chocolates to put in the box.

MATERIALS:

Materials may include:

- chocolates
- icing
- candy decorations
- boxes
- candy trays or tray liners
- ribbons, bows, and ties to decorate the box
- wrapping paper

GRADED ASSIGNMENT SUGGESTIONS:

Grading suggestions include:

- a reflection based on the experiences
- a project about candy and how it is made
- a research project about chocolate and how it is made, as well as where it is harvested

CATEGORIES: Decorating, Design, Food

VALENTINE'S DAY: Grab Your Hearts

Devise a tool that can help you grab pieces of candy from a distance. Fill your valentines candy bag as quickly as possible!

MISSION RULES AND RESTRICTIONS:

1. Design a device that allows you to pick up candy without touching it directly.

2. You cannot simply use two objects, like toothpicks, and pinch the candy to pick it up. You must actually assemble a tool or device.

3. Your device should be able to pick up a variety of sizes and shapes of candy.

4. Fill your valentines cup or bag faster than other teams.

MATERIALS:

Materials may include:

• popsicle sticks

• plastic straws

• rubber bands

• toothpicks

• tape

• glue

• index cards

• a variety of shapes and weights of candy

• small paper bags or cups

GRADED ASSIGNMENT SUGGESTIONS:

Grading suggestions include:

• a write-up of ideas, including blueprints

• grade results based upon if the device actually works, and, if so, how successfully?

• a reflection based on the experiences

• a project comparing different types of kitchen, garden, and garage tools, possibly classifying and sorting them by usage

CATEGORIES: Accuracy, Design, Task Completion

VALENTINE'S DAY: Heart-Shaped Air Mail

Build a paper airplane that can carry a valentine as far as possible.

MISSION RULES AND RESTRICTIONS:

1. Using the given materials, design a paper airplane that can hold either candy hearts or a valentines card as far as possible.

2. Fold and design carefully to fulfill the task.

3. OPTION: Make this an accuracy project instead. Planes need to land on a tabletop or inside a hula hoop.

MATERIALS:

Materials may include:

• Paper

• rulers

• candy hearts or valentines cards

GRADED ASSIGNMENT SUGGESTIONS:

Grading suggestions include:

• a write-up of ideas, including blueprints

• grade results based upon if the device actually works, and, if so, how successfully?

• a reflection based on the experiences

• a project about air mail or the history of the postal service.

CATEGORIES: Accuracy, Design, Distance, Fliers, Motion, Planes

VALENTINE'S DAY: Mail Sorters

One of the most famous symbols of Valentine's Day is the little cards that people hand out to each other. Your mission is to design a project that can sort valentines by size or shape.

MISSION RULES AND RESTRICTIONS:

1. Design a mail sorting device.

2. Your device may just look like a tray that your drop mail into or an actual mailbox.

3. Valentines of different sizes and shapes should be sorted into different bins.

4. Redesign as necessary to get the best results.

MATERIALS:

Materials may include:

• 3-10 valentines cards of at least 2 different sizes or shapes.

• cardboard

• tape or glue

• plastic straws

GRADED ASSIGNMENT SUGGESTIONS:

Grading suggestions include:

• a write-up of ideas, including blueprints

• grade results based upon if the device actually works, and, if so, how successfully?

• a reflection based on the experiences

• a project about valentines cards, mailboxes, mail sorting, or postal history

CATEGORIES: Design, Shapes, Sorting, Task Completion

VALENTINE'S DAY: Tower of Hearts

Build the tallest tower possible from toothpicks and gummy hearts.

MISSION RULES AND RESTRICTIONS:

1. Teams should be of 2-3 people.

2. Using gummy hearts or gumdrops and toothpicks, assembled and build the tallest tower possible.

3. Each team gets a set amount of materials or a set amount of time.

4. 12 candies and 12-20 toothpicks is a good starting amount, if you are going by limited materials.

5. Otherwise, limit the time to 5, 10, or 15 minutes, and do not limit the materials.

MATERIALS:

Materials may include:

• gummy hearts, gumdrops, or other sticky heart-shaped candy (heart-shaped marshmallows also work)

• toothpicks

• rulers

• timers

GRADED ASSIGNMENT SUGGESTIONS:

Grading suggestions include:

• a write-up of ideas, including blueprints

• grade results based upon if the device actually works, and, if so, how successfully?

• a reflection based on the experiences

• a project about towers, particularly researching a famous tower from around the world

CATEGORIES: Design, Engineering, Height, Measurement, Towers

HOLIDAY:

MISSION RULES AND RESTRICTIONS:

MATERIALS:

GRADED ASSIGNMENT SUGGESTIONS:

CATEGORIES:

AUTHOR: Andrew Frinkle

BRIEF:

A quick look at the author of over 100 titles!

ABOUT THE AUTHOR:

1. Over 10 years of experience in the teaching field with a specialization in math and science education in elementary and middle schools.

2. Award Nominated for teacher of the year.

3. Winner of the Karen Pelz Writing Award for short fiction.

4. Author of over 20 novels and books of fiction under the pen name Velerion Damarke and over 80 educational titles.

5. Black Belt in the Korean Sword Martial Art Geomdo.

6. Owner/Operator of MediaStream Press.

SNAZZY PHOTO:

VISIT ME:

www.MediaStreamPress.com

- www.50STEMLabs.com

- www.AndrewFrinkle.com

- www.common-core-assessments.com

- www.littlelearninglabs.com

- www.spellwars.weebly.com

- www.underspace.org

EMAIL ME:

mediastreampress@gmail.com

CATEGORIES: Hands-On, Labs, Math, Measurement, Physics, Science, STEM

50 STEM LABS SERIES

50 LEARNING LABS SERIES

MEDIASTREAM PRESS

MediaStream Press offers over 120 fun and innovative books and games to help educate. Learn more at: www.MediaStreamPress.com or search and buy directly at: www.amazon.com/author/andrewfrinkle.